10|98

21.⁰⁰

# LIVES INTERTWINED
## RELATIONSHIPS BETWEEN PLANTS AND ANIMALS

### ALLEN M. YOUNG

A FIRST BOOK

**FRANKLIN WATTS**
A Division of Grolier Publishing
New York   London   Hong Kong   Sydney
Danbury, Connecticut

# DEDICATION

*Many thanks to my parents, George and Margaret Young, and Marion Cormier who allowed and encouraged me to scour the fields and woodlands in search of insect discoveries during my youth.* ● ● ● ● ● ● ● ● ● ● ● ● ● ● ● ● ● ● ● ● ● ● ● ● ● ● ● ● ● ● ● ● ● ● ● ● ● ● ● ● ● ● ● ● ●

Cover and interior design by Molly Heron
Photographs ©: Allen M. Young: Cover, pp. 4, 6, 12, 18, 22, 24, 27, 30, 32, 34, 36, 43, 46, 48, 49, 52, 54; Photo Resaerchers: p. 15 (Renee Lynn); Visuals Unlimited: pp. 9, 16 (Jo Earney), 20 (G. Prance), 38 (Kjell B. Sandved), 40 (Joe McDonald).
Author photo ©: Linda Port
Illustration by Maud Kernan

Library of Congress Cataloging-in-Publication Data

Young, Allen M.
Lives intertwined: relationships between plants and animals/ by Allen M.Young
p.cm.—(A First book)
Includes bibliographical references and index.

Summary: Describes the interdependence of plants and animals in a Central American rain forest, focusing on the Morpho butterfly and the Mucuna vine.

ISBN 0-531-20251-8
1. Animal-plant relationships—Central America—Juvenile literature.
2. Rain forest ecology—Central America—Juvenile literature. 3. Morpho peleides—Ecology—Central America—Juvenile literature. 4. Mucuna Ecology—Central America—Juvenile literature. [Animal-plant relationships. 2. Rain forest ecology. 3. Ecology. 4.Morpho. 5. Butterflies. 6. Mucuna. Plants.]   I. Title
II. Series.
QH549.5.Y68   1996
574.5'24—dc20                                                   96-4923 CIP  AC

## ACKNOWLEDGMENTS

I would like to thank Gretchen Will Mayo who, upon hearing one of my slide presentations on insects and plants, recommended that I write a book for Franklin Watts.

Many individuals and institutions provided me with opportunities to study insects in the temperate and tropical settings. These include Heinz Meng, Arnold Nemerofsky, James Gray, Thomas Park, Monte Lloyd, Lynn Throckmorton, Daniel Janzen, Lincoln Brower, E. O. Wilson, Thomas Moore, Murray Blum, Robert Hunter, the State University of New York at New Paltz, the University of Chicago, the Organization for Tropical Studies, Inc., Lawrence University, and the Milwaukee Public Museum. Much of the work was funded by grants from the National Science Foundation, Friends of the Milwaukee Public Museum, and the American Cocoa Research Institute.

I am also indebted to Pat Manning of the Milwaukee Public Museum and to my editors at Franklin Watts for their assistance with the manuscript.

# CONTENTS

# CHAPTER 1
## ANIMALS NEED PLANTS

Dawn is breaking over the tropical rain forest of Central America. The animals that hunt in the darkness of night have just fallen asleep among the branches of the tall trees. The animals that feed during the day are just beginning to awaken. For a few minutes, the rain forest is a safe place.

A young red-and-yellow caterpillar with a large, bristle-covered head slowly crawls out from its hiding place, the underside of a Mucuna leaf, and begins to eat. After about 15 minutes, the Morpho caterpillar

*The Morpho caterpillar eats Mucuna leaves for several months until it is fully grown. Mucuna is one of the few species of rain forest plant on which the Morpho butterfly lays its eggs.*

crawls back under its leaf and remains perfectly still all day long. As the sun begins to set, the caterpillar crawls back out and feeds for another 15 minutes.

By feeding only at dawn and dusk, Morpho caterpillars are able to avoid being spotted by their *predators* (the animals that would like to eat them).

Although several kinds of Morpho live in the tropical rain forests of Central America, our story concerns *Morpho peleides*, the most common species of Morpho. Throughout this book, *Morpho peleides* will be referred to simply as Morpho.

The Morpho caterpillar shares the Mucuna plant with many different kinds of plant-eating insects. These insects depend on the Mucuna plant to survive. Like many insects in the tropical rain forest, the Morpho caterpillar eats only a few, select kinds of plants. If something happened to the Mucuna population, Morpho caterpillars would suffer, too.

When you think of the relationship between plants and animals, you probably think of an example like that of the Morpho caterpillar and the Mucuna plant. In this relationship, the animal needs the plant.

Plants provide Morpho caterpillars, and most other animals, with the food and shelter they need to survive. Plants also provide the oxygen that animals breathe. All over the world, animals need plants to survive. You are one of these animals.

Think of all the different types of food that you eat. Many of these foods come from plants. Vegetables, fruits, bread made from wheat or other grains, rice, and nuts all come from plants. Almost all of the animals that you eat fed on plants while they were alive. You couldn't survive without plants, either.

You may already know how important plants are to animals. But did you ever think about how much plants need animals? Without animals, most plants would not be able to *reproduce*. Plants need animals just as much as animals need plants.

*All of these foods come from plants. Animals, including humans, rely on plants for food to eat and oxygen to breathe.*

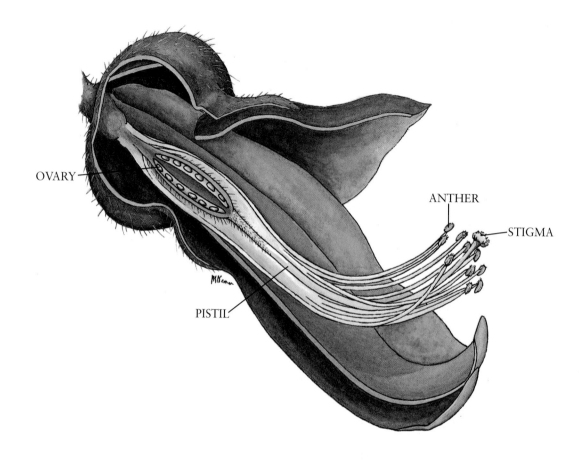

OVARY

ANTHER

STIGMA

PISTIL

*During pollination, pollen is transferred from the anther of one flower to the stigma of the same flower or a different flower. The pollen moves down a tube within the pistil. When it reaches the ovary, it fertilizes an egg.*

# CHAPTER 2
## PLANTS NEED ANIMALS

It's close to midnight. The tropical rain forest is alive with the flutter of thousands of bats in search of their favorite foods. Some species of bats are hunting insects, frogs, or fish. Others are looking for sweet, sugary nectar.

A few hours earlier the tiny flowers of the Mucuna plant began to blossom. As the flowers opened, they emitted a musky scent that attracted bats roosting under the large, sturdy leaves of nearby palm trees.

The hungry bats lap up the nectar as they crawl among the clusters of flowers that hang down from the long Mucuna stems. As they move from flower to flower, the bats unknowingly pick up *pollen* (the male sex cells of green plants).

Some of the Mucuna pollen is deposited on other flowers of the same plant as the bats continue to feed.

All Mucuna flowers have both male and female parts. The bats may also carry pollen from one Mucuna plant to another.

*Pollination* occurs when a flower's pollen is transferred from the *anthers* (male parts) onto the *stigma* (female part) of the same flower, another flower on the same plant, or a flower on a different Mucuna plant. Pollen grains, stripped of their outer coats, move down a special tube inside the flower to *fertilize* egg cells in the *ovary*.

If fertilization is successful, seeds will develop and the ovary will become a fruit. Because the Mucuna is related to pea and bean plants, Mucuna fruit looks like a large brown pea pod. Each pod is about 12 inches (30 cm) long and is covered with a dense layer of tiny bristles. Clusters of pods dangle from tall, woody Mucuna vines.

When a pod ripens, it splits open lengthwise and several hard black seeds fall to the ground. Sometimes, entire pods drop off the vine. It may take months or

*Ripe Mucuna pods sometimes split open as they drop to the forest floor. Inside are tough blackish-brown seeds (called "ox-eyes").*

years for seeds to soften in the damp leaf mulch and sprout new vines.

Mucuna pollination could not occur without bats. Plants all over the world need insects, birds, and other animals to help with pollination.

Pollination is a partnership between animals and plants. It is a partnership because both the animal pollinator and the plant being pollinated benefit. Pollinators usually receive a "reward" from the flower during pollination. The bats that pollinate Mucuna receive energy-rich nectar.

The plants benefit too. Without pollination, fertilization cannot occur and new seeds cannot be produced. Because bats often travel long distances through the rain forest, they may *cross-pollinate* plants that are miles apart. Cross-pollination can introduce new characterics that make the plants stronger.

If you look at all of the plants growing in a local garden or field, you will notice that their flowers come in a wide variety of shapes, colors, and sizes.

*As you can see by looking at the geranium, columbine, and foxglove shown here, flowers come in a variety of sizes, colors, shapes, and scents.*

*Humans enjoy spitting out watermelon seeds on a hot summer day. In nature, many animals, especially birds and mammals, routinely eat fruits and spit out or defecate seeds through the forest. These seeds eventually sprout into new plants.*

Each flower also has a unique scent. There is a reason for this.

Each kind of flower attracts only certain kinds of animal pollinators. Mucuna plants have a strong scent, but are not very colorful. This is because bats have a strong sense of smell, but very poor vision.

Flowers are not the only plant parts that attract animals. Many animals are also attracted to seeds and fruits. This is because many plants rely on animals to *disperse* their seeds.

Animals—including humans—eat the fruits and inadvertently carry the seeds to new locations. Have you ever spit out seeds as you ate a slice of watermelon? If so, you dispersed the seeds for the plant.

If you returned to the spot where you spit out the seeds months or years later, you might see a plant growing there—or you might not. One of the reasons that fruits have so many seeds is because very few actually grow into plants. Seeds that have been carried to a new site may be more likely to grow because they will not compete with as many other seeds for nutrients and space in which to grow.

A wide variety of animals eat fruits and scatter seeds in tropical rain forests. As a result, it is unusual to find two or more trees of the same species growing next to each other.

In the rain forest, many *species* of fruit-eating bats

and birds help to fill in gaps created when giant trees die and fall over. How do birds and bats do this? They drop the seeds from different plants into these openings. Then, these new plants provide food and shelter for many other animals.

## How Mucuna Grows

The Mucuna plant is a woody vine with oblong, dark-green leaves. As it stretches toward the sunlight, it wraps itself around trees. It must reach the *canopy* (the top of the rain forest) before it can produce flowers.

A variety of medium-sized mammals, including tapir, deer, peccaries, and monkeys, roam Central American rain forests in search of fruits, including Mucuna seed pods. These animals help Mucuna plants by carrying the seeds to new locations. They pluck ripe Mucuna pods off the plant or pick them up off the ground. As the animals eat the pods, digestive juices in their mouths soften the hard Mucuna seeds.

*Mucuna vines usually sprawl out over the rain forest canopy. Note the machete used to give a scale of size.*

19

The tapir, shown here in a Brazilian rain forest, eats plants and fruit. It scatters seeds that pass though its digestive tract or that are dropped as it feeds.

As a result, the seeds can sprout as soon as they pass through the animals' *digestive systems.*

Mucuna seeds are also scattered by squirrels and agouti, two types of small rodents. These animals often collect the seeds, transport them to new locations, and then forget where they have left the seeds. If squirrels and agouti do return to their stash, they discover that they cannot break open the large, hard seeds. The Mucuna seeds scattered by these small mammals must wait for the moisture and chemicals found in leaf mulch to soften the seeds enough for *germination* to occur.

As these seeds soften over the course of months or even years, the forest is changing. Dozens of trees are growing. By the time the Mucuna seeds are ready to break open, the forest floor has become a dark place.

Without light, the Mucuna seeds cannot sprout. They must wait for a giant tree to die and fall, creating an opening in the forest.

When a tree does fall, young Mucuna vines, and many other plants, sprout in the new opening. Although many of the shoots are eaten by hungry animals, some survive.

The young Mucuna vine grows quickly. By the time the vine reaches the canopy, it is a thick, woody, ropelike *liana*. Its leaves sprawl over the branches and foliage of a number of nearby trees.

When a vine finally reaches the canopy, it can produce flowers. Hopefully, the flowers will be pollinated by bats and produce more seeds.

In the tropical rain forest, most green plants have a unique set of animal pollinators and seed dispersers. This is one of the reasons that tropical rain forests

contain a greater variety of plant and animal species than any other type of ecosystem.

The size, shape, color, pollen, nectar, fragrance, and position of each flower attracts specific pollinators. Similarly, the size, color, shape, and taste of each fruit attracts a specific seed disperser. Without these animals, the plants would not be able to reproduce.

Plants need animals as much as animals need plants. In the natural world, every creature depends on many others. Without these interrelationships, no living thing could survive or *perpetuate* its own species.

⬤ ⬤ ⬤ ⬤ ⬤ ⬤ ⬤ ⬤ ⬤ ⬤ ⬤ ⬤ ⬤ ⬤ ⬤ ⬤ ⬤ ⬤ ⬤ ⬤ ⬤

*Central American rain forests contain many species of large, woody vines called lianas. They may stretch 100 to 200 feet (30 to 60 m) above the ground. Some lianas sprout aerial roots that reach down to the forest floor.*

# CHAPTER 3
## ANIMALS PROTECT PLANTS

For the Mucuna vine to reach the canopy, it must wrap itself around the trees in the forest. As a Mucuna vine grows, it often chokes these trees and stunts their growth.

Most trees cannot do anything to prevent injury caused by the Mucuna and other vines. The Cecropia tree is one exception.

Over thousands of years, it has developed a *symbiotic relationship* with Azteca ants. The Cecropia

*The Cecropia tree grows in light gaps (small clearings) along the borders of the Central American rain forest. Although trees and shrubs in these exposed areas are often smothered by vines, the symbiotic ant inhabitants protect the Cecropia by cutting back vines that attempt to grow up the trunk and cover the tree.*

provides the ants with food and shelter. In return, the ants attack Mucuna and other vines that threaten the tree's ability to grow.

The tall, slender trunk of the Cecropia tree is hollow and filled with millions of Azteca ants. The hollow trunk provides the Azteca ant with a home and food. The tree has special "food bodies" called trichilia. These trichilia produce tiny droplets of fluid that are rich in sugars, fats, and proteins. The droplets are collected and eaten by the ants and their *brood*.

A fertilized Azteca queen ant must locate a Cecropia tree seedling before she can establish her own colony. Once she finds the tree, she chews a hole in the soft stem and lays her eggs. When these eggs hatch, she will have an army of worker ants. The ant colony grows and flourishes as the little Cecropia tree grows.

When Mucuna or other types of vines wrap themselves around a Cecropia, the ants living inside the tree use their sharp jaws to clip away the vines. When Azteca ants are absent or die off, the Cecropia tree also dies.

*Azteca ants nest inside the hollow Cecropia trunk. The ants feed on sugar droplets secreted by the tree.*

# CHAPTER 4
## DON'T EAT ME

Azteca ants are not able to protect Cecropia trees from all potential dangers. A variety of *herbivores* feed on the leaves, fruits, and seeds of the Cecropia and all other plants. Herbivores are animals that eat living green plants.

Each kind of plant attracts a unique set of enemies. Over thousands of years, plants have developed strategies to reduce the amount of damage done by herbivores.

While developing on the vine, Mucuna seed pods are protected by sharp fine brown hairs. These hairs come off easily and stick to the lips and face of monkeys and other mammals that are after the soft "milk" seeds developing inside.

Despite such defense systems, no plant can protect itself against every possible attacker. Mucuna plants have no way to protect themselves against seed-eating weevils that lay rafts of eggs on the outer surface of Mucuna pods.

Morpho caterpillars, discussed in Chapter 1, also threaten the Mucuna plant's ability to grow and re-produce.

Animals have enemies, too. You have already learned that Morpho caterpillars feed for only a few minutes at sunrise and sunset to avoid their predators. Adult Morpho butterflies must protect themselves against birds and lizards.

Like plants, most animals have developed special ways of protecting themselves against their attackers. Morpho butterflies try to hide from predators by blending in with their surroundings.

The wings of Morpho butterflies are a beautiful, iridescent blue color on top.

This eye-catching metallic blue resembles the rain-bow-hued shimmer of a soap bubble or an oil slick. While the colors that you see on most butterfly wings are caused by pigments in the scales that cover the wings, the colors that you see on Morpho wings are produced by microscopic grooves and ridges along colorless scales.

Morpho scales, which are arranged like the shin-gles on a roof, reflect sunlight. As waves of light ap-proach these scales, they interfere with other waves

A *Morpho* basks in the morning sunshine in a rain forest clearing. Although the butterfly usually perches with its wings tightly shut, extending its wings helps it to warm up before a long flight.

that have been reflected. Some of the light waves cancel each other out, but the waves of blue light are intensified.

Insect-eating birds such as jacamars can easily spot Morpho butterflies and attack them in midair. These birds often descend on male Morphos as the butterflies travel along their morning patrol routes. The only protection a Morpho butterfly has against the jacamar is its ability to alter its flight pattern. As soon as a Morpho senses that one of its enemies is close by, it begins to fly in an unpredictable, zigzag pattern.

The challenge to snare the butterfly is made even more difficult by flash patterns produced by the butterfly's iridescent blue wings. One moment the wings appear to be a brilliant, shimmering blue. Suddenly, as the sun hits the butterfly's wings from a different angle, they begin to look deep purple or even black. One moment the Morpho seems an easy target, the next it has blended into the lush background of rain forest vegetation.

If the Morpho butterfly's bright blue wings were always visible, the Morpho wouldn't stand a chance against its predators. Because the blue coloring on the top of their wings would make them an easy target, these butterflies keep their wings tightly closed when they are not flying.

A *Morpho* butterfly perches on the trunk of a felled rain forest tree in Costa Rica as it feeds. The brown and gray patches on the undersides of its wings act as camouflage.

Instead of seeing the blue on the top of their wings, predators see the gray and brown on the undersides of their wings. These colors blend with the leaf mulch on the forest floor, making it difficult to see the butterflies at all. The Morpho butterflies can feed on rotting fungi without any fear of being spotted.

The undersides of Morpho wings also have "eye-spot" markings. Birds and lizards may mistake these markings for the real eyes of larger animals and be scared away.

Female Morpho wings are a little less brightly colored. In addition, females are much less likely than males to be spotted flitting along open creeks and paths in the bright morning sun.

Some butterflies have evolved chemical defense systems to avoid being eaten by their enemies.

The bright, gaudy colors and the thick, rubbery bodies of Parides butterflies advertise their distastefulness to potential predators. Parides butterflies taste bad because when they were caterpillars, they ate Archistolochias, a poisonous plant. This poison is stored in the body tissue of the adult butterflies. Parides caterpillars are one of the few insects that is not poisoned by Archistolochias.

Some rain forest birds are able to eat Parides caterpillars because they pluck the caterpillar's head off and pull out the digestive tract, which contains poisonous Archistolochias leaf fragments. The bird can then swallow the rest of the caterpillar.

Parides caterpillars have a second chemical weapon. If an ant bites one of these caterpillars, a

The bright orange osmeterium just behind the head of this Parides caterpillar is exposed only when the animal is threatened.

Y-shaped yellow or orange organ pops out from behind the caterpillar's head, and a very strong, disagreeable smell is given off. The attacker usually backs off when this happens. The pungent odor is made by the caterpillar's body and does not come from its food plant.

There are several species of Parides living in the tropical rain forests of Central America. Without examining them very closely, it is impossible to tell the difference between the species. Because the butterflies look alike, an animal that has eaten any one of these Parides butterflies will not eat another one. The predator will remember how terrible the first individual tasted. Scientists call this defensive strategy *Mullerian mimicry.*

Since the butterflies intermingle considerably, each species suffers fewer losses than if just one species was distasteful. Because their rubbery body protects them from being eaten quickly, the butterflies have enough time to give off a strong odor. When the attacker smells this odor, it drops the uninjured butterfly.

In another kind of protective association, called *Batesian mimicry,* distasteful insects such as Parides are models for tasty mimics such as the swallowtail butterfly *Papilio anchisiades.* Because the mimic closely resembles the model, attackers cannot tell

Which one doesn't belong? These mounted specimens represent several species of *Parides* butterflies and a swallowtail butterfly (lower right corner). Note the striking similarity (mimicry) in the color patterns, shape, and size of the wings.

which is which. As a result, many predators avoid both butterflies.

Even though many animals have developed elaborate strategies to protect themselves, no animal can avoid every predator. Virtually all kinds of caterpillars, including the Morpho caterpillar, are attacked by *parasitic* flies and wasps.

By feeding only at dawn and dusk, Morpho caterpillars can avoid being spotted and eaten by some of their predators. The caterpillars cannot hide from parasites, however. The parasites use their sense of smell to locate the caterpillars.

Once a parasite finds a Morpho caterpillar, it begins to feed on the living caterpillar's body. As soon as the parasite lands on the Morpho caterpillar, the caterpillar begins to thrash about violently. A small flap of fluid-filled tissue located on the caterpillar's underside expands and releases a strong odor, which may cause the parasite to end its attack. In most cases, however, the caterpillar dies a slow, cruel death.

Some parasites of Morpho caterpillars deposit their eggs on Mucuna leaves. The caterpillars gobble up these eggs when they feed on the leaves. The eggs hatch inside the body, and the *maggots* (the larval stage of flies) begin to feed on the caterpillar.

Other parasites lay eggs directly on the host insects.

*The caterpillar of a sphinx moth in a South American rain forest is covered with the tiny, white silken cocoons of a braconid wasp. Braconid wasps parasitize the caterpillars of many species of butterflies and moths. The tiny wasp lays its eggs on the caterpillar's body, and the hatchling larvae burrow into the flesh to feed. When fully grown, they emerge from the host caterpillar and build cocoons.*

When these eggs hatch, the maggots must burrow into the caterpillar's body before they can feed.

While parasites are bad news for Morpho caterpillars, they are good news for Mucuna plants. Without parasitic wasps and flies, the caterpillars would eat so many leaves that the plant would die.

If caterpillars were allowed to kill too many Mucuna plants, the plants could not produce new seeds. Without new seeds, there would be no new Mucuna plants for future generations of Morpho caterpillars to eat.

As you can see, the parasites play an important role in maintaining a delicately balanced relationship. Death is a natural part of life. Every living creature eventually dies, decays, and rots away.

# CHAPTER 5
## WHO EATS A JAGUAR?

D ead plants and animals do not just disappear. In nature nothing is wasted, and creatures are not wasteful. Energy is unlocked by animals as they eat living plants and each other. As the food is converted from one form to another, energy is carried up the food chain.

So what happens at the top of the food chain? When a harpy eagle, or a boa constrictor, or a jaguar dies, its body is broken down by *decomposers*.

Decomposers are organisms that feed on dead or

*In the Central American rain forest, the jaguar is at the top of the food chain. When a jaguar dies, its body is broken down by decomposers. As a result, the nutrients stored in the jaguar's body are released into the soil.*

dying plant materials, including fallen fruits and seeds, dead animals, and waste matter. When these materials are broken down by decomposers, the nutrients that they contain are released back into the soil.

In tropical regions, vultures often begin the process of decomposition by feeding on the bodies of dead snakes, birds, and mammals. Flies lay their eggs on the rotting flesh exposed by vultures. When the fly eggs hatch, the maggots that emerge eat the rotting flesh. This rotting flesh is also attacked by bacteria and fungi.

Every day in the rain forest, vast numbers of dead leaves, twigs, branches, seeds, and fruits drop to the ground, creating a thick bed of mulch. Sometimes, giant trees that have been struck by lightning or attacked by insects fall over, adding to the mulch. The rotting processes in the leaf mulch, started by the tons of tree vegetation that rains down from the canopy, provide the nutrients needed by sprouting seeds and young plants.

Many decomposers, like bacteria, are so small that you can't see them. Other decomposers—grubs, millipedes, and ants—are more familiar. Mushrooms, toadstools, and other types of fungi are decomposers, too.

In addition to breaking down dead matter, decomposers play another important role in maintaining

*These mushrooms or fungi break down decaying or dead plant materials, adding life-giving nutrients to the mulch and soil.*

life in the tropical rain forest. Like all other creatures, decomposers are eaten by predators.

Morpho butterflies are decomposers. They feed on dead animals, waste matter, and rotting fruit. Morpho butterflies are also predators of decomposers. One of their favorite foods is the juice found in the caps of sting horn fungi.

As a Morpho butterfly searches for breakfast among the moldy leaf mulch scattered on the forest floor, it picks up decaying *spores* (male sex cells of fungi) and bacteria. The spores cling to the butterfly's mouthparts and the bristles on its legs. When the butterfly flies off, it sprinkles the leaf mulch with these spores and bacteria. As a result, these decomposers are spread throughout the rain forest.

Like the bats that pollinate Mucuna flowers, Morpho butterflies play an important role in the reproductive cycle of fungi.

# CHAPTER 6
## A NEW GENERATION

I t is a warm, sun-drenched morning. A Morpho butterfly floats down through an opening in the rain forest canopy. As it weaves through scattered thick boughs of lush vines, it resembles a shimmering blue spark.

On this particular morning, the female butterfly is looking for a Mucuna plant. She has recently mated with a male Morpho and wants to lay her fertilized eggs. The male was lured to the female by the attractive blue color of her wings and a special scent given off by her body.

Out of the thousands of plants thriving here, the butterfly will choose a Mucuna as a home for her eggs. This is because Morpho caterpillars eat only a few select species of plants.

When the female Morpho spots the right kind of

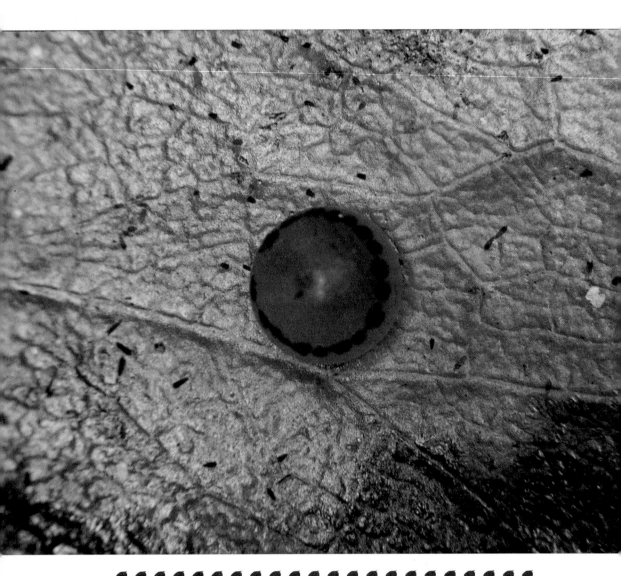

*The 1/16-inch (2-mm) diameter Morpho egg, with its translucent green color and reddish-brown ring, is difficult to spot on the upper side of a Mucuna leaf.*

leaf, she lands or holds onto it while fluttering her wings. She then taps the leaf surface with her antennae and front legs, picking up the scent through special hairs and pores. If the scent is correct, the butterfly places an egg on the leaf by touching the tip of her grayish body to the leaf surface to attach an egg. She then flies to another leaf and repeats the process.

The Morpho carefully places her eggs, one at a time, on the top sides of the oldest dark-green Mucuna leaves. Each egg is a tiny, translucent, green, dome-shaped object. It is well camouflaged against the leaf surface.

Soon, the tiny egg develops a dark reddish-brown ring, a sign that it is healthy and a caterpillar is being formed inside. Less than two weeks later, if the egg is not discovered and eaten by an ant or predatory fluid-sucking bug, it hatches into a red-and-yellow caterpillar with long bushy black hairs covering its head.

The baby caterpillar's jaws, called mandibles, are very strong, and after it eats the eggshell, the caterpillar begins to feed on the leaf. Morpho caterpillars anchor themselves in silken walkways as they crawl on the leaves and stems. This protects them from being knocked off in heavy rain and wind.

Many butterflies place eggs on young tender leaves and shoots, which are more nutritious and eas-

*Upon hatching, the ³⁄₁₆-inch (5-mm) Morpho caterpillar eats its own vacated eggshell before eating leaf tissue.*

ier to chew. But Morpho caterpillars start out life on the older tough leaves of Mucuna.

It takes about 100 days for a Morpho caterpillar to become fully grown. By this time, it has shed its outer layer (cuticle) several times and has eaten its fill of Mucuna leaves.

*The newly emerged adult Morpho butterfly, shown here still clinging to the chrysalis, cannot fly for several hours.*

The caterpillar forms a plump, teardrop-shaped green *chrysalis* suspended from a stem or twig by shedding its cuticle one last time. Although large and plump, the emerald-green chrysalis of Morpho blends well with the surrounding vegetation.

After about three weeks, the chrysalis begins to darken. The color changes as the green tissue and blood of the caterpillar is replaced by the black body and legs of the adult butterfly.

A few days later, in the early morning, a small tear line develops at the base of the chrysalis. A few moments later, the butterfly's antennae pop out, followed by its head. As the butterfly continues to push against the cuticle, its head, upper body, and front legs emerge. As soon as its front legs are freed, the butterfly begins to curl around the side of the cuticle. The butterfly grips the cuticle and pulls until the rest of its body pops out.

The newly hatched adult Morpho butterfly is completely defenseless. It hangs with its wings wrapped tightly around its body. Within a few hours, the force of gravity and the pressure of pumping blood allows the caterpillar to stretch its wings to their full span and fly off into the rain forest.

All that is left behind is the tough transparent cuticle, which is still nearly intact.

# CHAPTER 7

The Morpho cannot complete its life cycle without help from other organisms. As an adult, the Morpho needs food provided by dead animals, rotting fruit, and fungi. In turn the fungi rely on the Morpho butterfly to spread their spores. The Morpho caterpillar needs the Mucuna plant for food and shelter.

The Mucuna depends on other organisms, too. It needs bats to pollinate it, parasitic flies and wasps to control the population of herbivores, and decomposers to add nutrients to the soil.

Life in the tropical rain forest is a delicate balance. If the population of bats, parasites, or decomposers suddenly decreases, the population of Mucuna vines will be affected. So will the Morpho population. On the other hand, if the population of parasites

*This fully grown Morpho caterpillar is very different from younger ones. Although its earth-tone coloring helps it blend into its surroundings, many caterpillars fall victim to parasitic flies and wasps as well as carnivorous animals.*

suddenly increases, Mucunas might do well, but the population of Morphos will be severely affected.

The conditions on earth are always changing. Maintaining this delicate balance of interrelationships between plants and animals means that sometimes entire species will face *extinction*. Other species will *evolve* to take their place. All of these changes occur gradually. They take place over hundreds, or even thousands, of years.

In the past few decades, the tropical rain forests have experienced many changes. As they have been cleared and used for farming, the natural balance of species has been disrupted.

When there are massive, sudden losses of forest, plants and animals cannot survive because their food and shelter disappear very quickly. Many of the interrelationships that developed over thousands of years are being destroyed.

The destruction of tropical rain forests has also affected the wetland and woodland populations of North America. Many North American birds spend the cold winter months in Mexico, Central America, and South America.

As the tropical rain forests are destroyed, so are the insects and berries that these songbirds eat. As competition for food increases among the creatures that live in the tropical rain forests year round, there

is less and less food for migrants from North America. Fewer and fewer songbirds are returning to the United States every spring.

Many of these birds are also threatened by loss of their North American homes. As wooded areas are cleared to make room for housing developments and shopping malls, there are fewer places for these birds to build their nests and raise their young.

Protecting and preserving the earth's natural habitats provides humans with the chance to study and appreciate the many interrelationships between plants and animals, both at home and in the tropical rain forest.

*The tropical rain forest is being destroyed to make way for cattle pastures.*

# GLOSSARY

*anther*—part of the stamen of a flower that produces and releases pollen.

*Batesian mimicry*—a protective association in which a harmless species avoids predators because it resembles a poisonous species.

*brood*—offspring; an animal's young.

*canopy*—the area of the tropical rain forest that contains most of the treetops.

*chrysalis*—the stage of the life cycle between the caterpillar (larval stage) and butterfly (adult stage). During the period of transition, the insect resides in a tough cuticle structure.

*cross-pollinate*—to transfer pollen from the anther of

one flower to the stigma of a flower on a different plant.

*decomposer*—an organism that feeds on dying or dead organic matter. As decomposers break down matter, nutrients are released into the soil.

*digestive system*—the organ system in the body that breaks down food. Any substance that is not absorbed into the bloodstream is eliminated as feces.

*disperse*—to spread.

*ecosystem*—a group of organisms living in an environment with specific physical conditions.

*evolve*—to gradually change over millions of years to adapt to a particular environment.

*extinction*—the process by which a species dies out because it is unable to adapt to environmental changes.

*fertilize*—to unite an egg and sperm cell.

*germinate*—to sprout or start to grow.

*herbivore*—an animal that eats living green plants only.

*liana*—any one of a group of woody vines that grows in the tropical rain forest.

*maggot*—the larval stage of a fly.

*Mullerian mimicry*—a protective association in which a number of poisonous species resemble each other. Once a predator eats any one of these species, it will avoid all similar-looking species.

*ovary*—the part of a flower that contains ovules (where eggs are produced). It is at the base of the pistil. After fertilization, the ovary swells and becomes the fruit.

*parasitic*—harmful or destructive.

*perpetuate*—to cause something to last or continue for an indefinite period of time.

*pollen*—the male sex cells of green plants.

*pollination*—the process by which pollen is transferred from the anther (male part of the flower) to the stigma (female part of the flower). The pollen may be carried from one flower to another by insects, birds, bats, or the wind.

*predator*—an animal that survives by eating another animal.

*reproduce*—to create a new generation of a species.

*species*—a group of organisms that produce viable offspring when they mate.

*spores*—the male sex cells of fungi.

*stigma*—the tip of the pistil; the part of a flower upon which pollen is deposited.

*symbiotic relationship (mutualism)*—a specialized partnership in which two unrelated organisms live together and benefit one another. In some cases, one organism cannot survive without the other.

# FOR FURTHER READING

Brewer, J., and D. Winter. *Butterflies and Moths*. New York: Prentice Hall, 1986.

Durrell, G. *The Amateur Naturalist*. New York: Alfred A. Knopf, 1983.

Forsyth, A., and K. Miyata. *Tropical Nature*. New York: Charles Scribner's Sons, 1984.

Leahy, C. *Peterson First Guides—Insects*. Boston: Houghton Mifflin Co., 1987.

Lingelbach, J. *Hands-on Nature*. Woodstock, Vt.: Vermont Institute of Natural Science, 1986.

New, T. R. *Butterfly Conservation*. New York: Oxford University Press, 1991.

Pyle, R. M. *Handbook for Butterfly Watchers*. Boston: Houghton Mifflin, 1984.

Roth, C. *The Amateur Naturalist.* New York: Franklin Watts, 1993.

Snodgrass, R. E. *Insects—Their Ways and Means of Living,* 2nd ed. New York: Dover Publications, 1967.

Young, A. M. *Sarapiqui Chronicle: A Naturalist in Costa Rica.* Washington, D.C.: Smithsonian Institution Press, 1991.

Zim, H. S., and C. Cottam. *A Golden Guide—Insects,* 2nd ed. New York: Western Publishing, 1987.

# INDEX

# ABOUT THE AUTHOR

Allen M. Young grew up in a suburb of New York City. After graduating from the State University of New York at New Paltz and receiving a Ph.D. in zoology from the University of Chicago, he conducted research in the tropical rain forests of Central America for more than 25 years. During his trips to the rain forest, he took many of the photographs that appear in this book. Dr. Young is the author of more than 200 articles and has written five adult nonfiction books. He is currently the curator of zoology and vice president of collections at the Milwaukee Public Museum in Milwaukee, Wisconsin.